Tap Into Renewable Energie Sources

Renewable Energie Sources In The UK
Albert Sicard

Copyright © 2013 by Albert Sicard. All rights reserved worldwide. part of this publication may be replicated, redistributed, or given away in any form without the prior written consent of the author or the terms relayed to you herein.

LEGAL NOTICE: The Publisher has strived to be as accurate and complete as possible in the creation of this report, notwithstanding the fact that he does not warrant or represent at any time that the contents within are accurate due to the rapidly changing nature of the Internet.

While all attempts have been made to verify information provided in this publication, the Publisher assumes no responsibility for errors, omissions, or contrary interpretation of the subject matter herein. Any perceived slights of specific persons, peoples, or organizations are unintentional.

In practical advice books, like anything else in life, there are no guarantees of income made. Readers are cautioned to reply on their own judgment about their individual circumstances to act accordingly.

This book is not intended for use as a source of legal, business, accounting or financial advice. All readers are advised to seek services of competent professionals in legal, business, accounting, and finance field.

Table of Contents

Table of Contents 3
My Favorite Quote 5
Introduction 6
Why Should I Care? 9
Renewable Energy - What Is It? 13
What Types Of Reneable Energy Is Available? 16
Option 1: BioFuel 19
Option 2: Hydropower 23
Option 3: Solar Power 28
Pros & Cons Of An Isis Free Solar System 34
Option 4: Wind Power 36
Our Future With Renewable Energy? 42
100 Energy Tips You Can Apply Today 47
Conclusion 69
Resources 73

My Favorite Quote

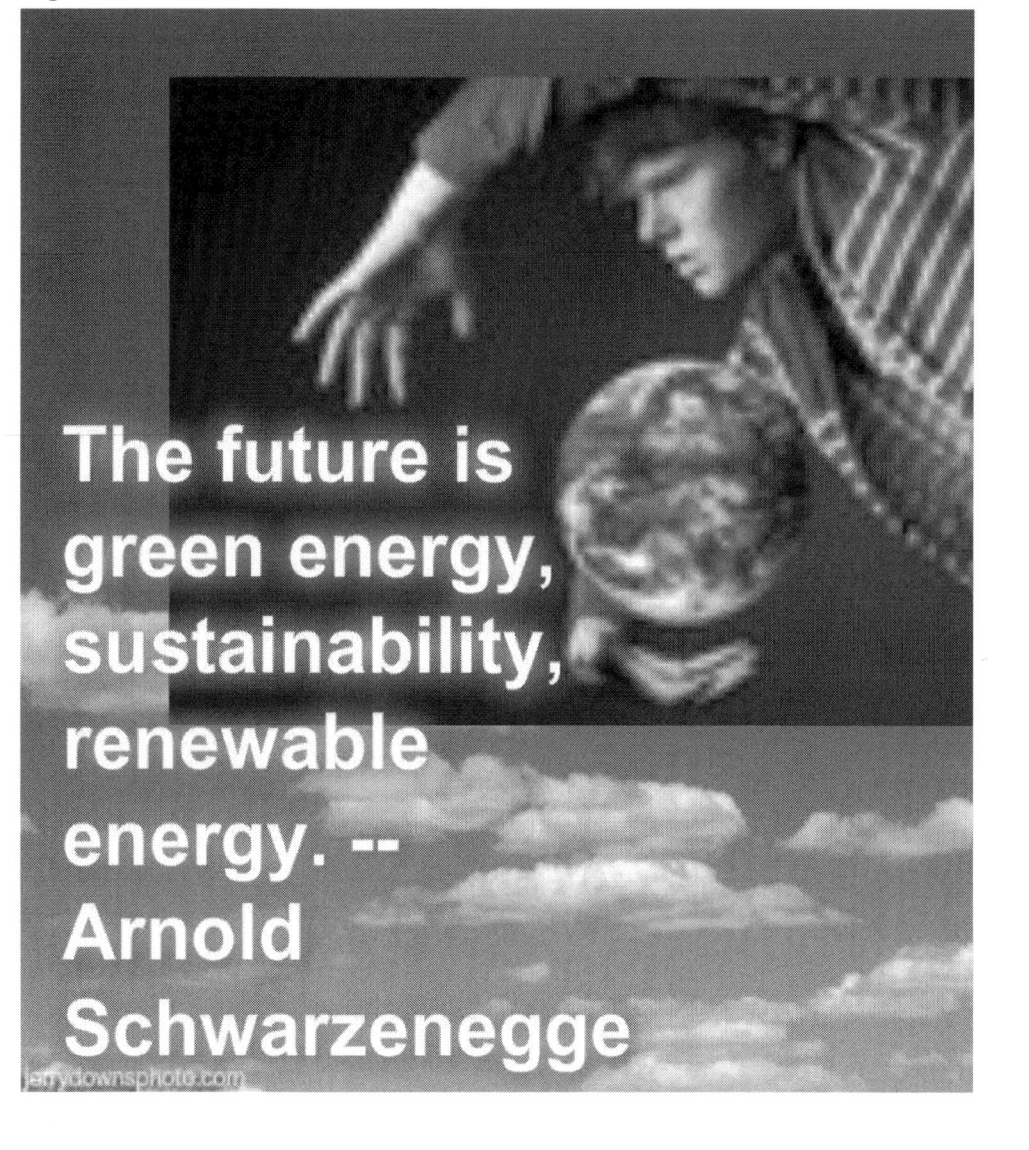

The future is green energy, sustainability, renewable energy. -- Arnold Schwarzenegge

Introduction

Do you know how to tap into renewable energy? If you want to learn how to harness the power of renewable, green and clean air for yourself and your family this powerful information is for you.

If you think that sustainable living is cool or if you are already living a sustainable life than you might want to tap into the power of renewable energy.

Renewable energy technologies are getting cheaper, through technological change and through the benefits of mass production and market competition.

A 2011 IEA report said: "A portfolio of renewable energy technologies is becoming cost-competitive in an increasingly broad range of circumstances, in some cases providing investment opportunities without the need for specific economic support," and added that "cost reductions in critical technologies, such as wind and solar, are set to continue."

Hydro-electricity and geothermal electricity produced at favourable sites are now the cheapest way to generate electricity. Renewable energy costs continue to drop, and the levelised cost of electricity (LCOE) is declining for wind power, solar photovoltaic (PV), concentrated solar power (CSP) and some biomass technologies.

Renewable energy is also the most economic solution for new grid-connected capacity in areas with good resources. As the cost of renewable power falls, the scope of economically viable applications increases. Renewable technologies are now often the most economic solution for new generating capacity. Where "oil-fired generation is the predominant power generation source (e.g. on islands, off-grid and in some countries) a lower-cost renewable solution almost always exists today.

Renewable energy is energy that comes from resources which are continually replenished such as sunlight, wind, rain, tides, waves and geothermal heat. About 16% of global final energy consumption comes from renewable resources, with 10% of all energy from traditional biomass, mainly used for heating, and 3.4% from hydroelectricity. New renewables (small hydro, modern biomass, wind, solar, geothermal, and biofuels) accounted for another 3% and are growing very rapidly. The share of renewables in electricity generation is around 19%, with 16% of electricity coming from hydroelectricity and 3% from new renewables.

100% renewable energy is where the trend goes in the future and you should be heading towards the 100% renewable energy movement, too.

Growth of wind and solar power wind, water and solar technologies can provide 100 per cent of the world's energy, eliminating all fossil fuels.

An expert about renewable energy advocates a "smart mix" of renewable energy sources to reliably meet electricity demand:

"Because the wind blows during stormy conditions when the sun does not shine and the sun often shines on calm days with little wind, combining wind and solar can go a long way toward meeting demand, especially when geothermal provides a steady base and hydroelectric can be called on to fill in the gaps."

Other renewable energy technologies are still under development, and include cellulosic ethanol, hot-dry-rock geothermal power, and ocean energy. These technologies are not yet widely demonstrated or have limited commercialization. Many are on the horizon and may have potential comparable to other renewable energy technologies, but still depend on attracting sufficient attention and research, development and demonstration.

This book shows you all the options that you have with renewable energy. It shows you what it is and what types of renewable energy is available to you, and gives you an outlook into the future, too. Finally, it supplies you with 100 usable and applicable Energy Tips that you and your family can get started today in order to save energy today and wake up to a brighter tomorrow with less bills to pay for energy.

Remember everything starts with commitment and I hope your commitment is to start learning about all your options with renewable energy.

My commitment is to educate you about all aspect of renewable energy and how you as a busy, modern and mobil world citizen can tap into the power of it and benefit from it in a brand new way. Once you understand all these interesting and fascinating aspects that go into renewable energy, I hope that your commitment and goal is going to take action and to approach renewable energy with this brand new knowledge in mind!

The book is not expensive and it is a very fascinating, interesting, easy and quick read so make sure you pick up your copy today so that you, too, can protect your family today!

It is true, renewable energy usage has grown much faster than anyone anticipated and you owe it to yourself and your family to tap into this powerful information of renewable energy so that you can survive in times of catastrophes and that you are able to live a sustainable lifestyle that secures your life and that of your family!

Welcome to a brand new green and clean world with renewable energy!

Why Should I Care?

We have come along way in developing societies that have electricity and the power necessary to fuel vehicles and for industry to be successful. All of these efforts though rely upon the use of energy source that comes from fossil fuels.

They are found in the ground and have to be processed in order for us to have that fuel and that electricity. They are known as coal, natural gas, and fuel. We rely on them way too much for our own good and that is why change is so important.

The problem though is that our dependence on it continues to grow. As more people are upon the Earth than ever before we are using more every single day. People are living longer too due to advances in health care. We are certainly a society dependent upon our electronic gadgets as well.

While those are all good things for us to be happy about, the fact that we are depleting the fossil fuel available isn't. This type of energy source isn't one that we will be able to replace. When it is gone it is gone and that is the reality of the situation.

It won't all disappear during our lifetimes, but it is going to pose a problem for future generations. While efforts can be made to converse fossil fuels, eliminating enough of the use in order to really make a difference isn't going to occur unless we take a close look at some alternative methods.

Society isn't going to go back to using horses and carriages for transportation. They also aren't going to go back to lighting their homes with candles at night. With the computer use around us in homes and for business it isn't even practical to suggest we stop using the electricity that is necessary to allow them to operate.

We can often take for granted just turning the key and our car starts, turning on the thermostat to have heating or cooling in our home, and flipping a switch to give us the lighting we need in any room. Some people are also selfish as they don't care what is going to happen for future generations as long as they have what they need right now.

Luckily, the majority of the population doesn't think that way. They aren't out to use everything they can without looking back. The problem though is that they often don't realize what they are using could be a problem down the road. Even if they do, they may not realize that they have some other options they can try to implement.

Learning about the various types of renewable energy is a great way to get a person thinking about changes they can implement. There have been some significant efforts made in this area but there is still much more than needs to be taking place. Instead of being afraid of what is unknown to you, do your best to learn the basics of all the renewable energy sources possible.

The government of the United States has gotten involved in promoting renewable energy sources as well. They offer some great financial incentives for homes and businesses to you them. Even so, there is sometimes a high overhead to get everything in place. This can prevent many people from being a part of saving our natural resources even when they really would like to.

It is estimated that about 13% of our current energy is the result of renewable energy. With the money to cover the expenses, advanced technology, and a desire by society to continue using them we can see that percentage significantly increase. There are plenty of benefits to renewable energy too such as not harming the environment with pollutants.

If you are asking yourself why we don't just turn to them now the answer isn't that simple. In a nutshell there is still a great deal of research that needs to be completed. There is also the high cost to contend with as well as various disadvantages with each of the types of renewable energy.

If you keep on reading though you will get to this information as well. Then it will make sense as far and the big picture of what we currently get from renewable energy, what the limitations are, and what we can expect into the future. This will help you to understand the benefits as well as the drawbacks of the situation more clearly.

Renewable Energy - What Is It?

Fossil fuels are used to make energy we use, but once they are gone we will never get more of them. Coal, oil, and natural gas all fall into this category. They are used all over the place in high amounts so you may not realize that they are in limited supply.

These forms of energy have been used mainly because they are affordable and they don't take up very much room to incorporate. They can be transported anywhere they are needed as well. With natural resources there have to be certain elements in place in order to take advantage of them.

The concept of renewable energy embraces the ability to use the resources we naturally have, but that we will never run out of. This way we can continue to have all the benefits we want without destroying the Earth.

We also won't be preventing future generations from having the chance to future grow and evolve beyond what we were able to see take place in our own lifetime. This process involves taking these types of natural resources and turning them into a product we can use for power.

That means a great deal of information and technology has to be collected and evaluated. Many of these methods though continue to see advances in the designs and processing which will result in them being even more valuable in the future than they are right now.

Most will agree that renewable energy sources are better for the environment. The burning of fossil fuels including gasoline and coal isn't good for the environment. These natural resources will allow us to save resources and at the same time to live in a cleaner environment than we have now.

Many believe it isn't practical or safe to depend only upon these types of resources though. That is because the sun doesn't always shine so the energy can't be collected. There are many places where the sun is blocked for days due to the changes in the seasons.

You can't predict how much wind will be produced or how much power can be taken from the water. It will vary significantly but there is no reason why we can't rely upon these renewable energy sources as the primary providers. We can then depend on fossil fuels as back up so we never have to go without the energy we want to use.

Chances are you have heard about the various forms of renewable energy but not in detail. Keep on reading and you will get all the information you need including the pros and cons of each type. You will find each of them does offer some hope for the future though as far as reducing our dependency on those resources which we can't replenish.

What Types Of Reneable Energy Is Available?

When you think about the natural things around us a couple of things come to mind. First, you have the sun that continues to shine brightly in the sky day after day. It gives off a great deal of heat which can be converted into energy. The sun shines brightly some days and then is covered with clouds other days. So the amount of energy you can collect each day is going to vary.

Next, you have water which covers the majority of the surface of the Earth. There is also the additional moisture and rainfall that can be collected as time goes by. There is energy found in the water as it moves along and this can be converted into energy at hydropower plants.

Even if it is just barely there on certain days, you also have the wind. In some areas it is extremely windy all the time. In order for the equipment used to create wind power to be worth the cost it must be blowing most days at a speed of at least 15 miles per hour.

Most people view that as a nuisance but they don't realize the full potential of it. They aren't really aware that the wind that is all around you can be used to create renewable energy. It is also very clean for the environment so you don't have to worry about negative effects from it.

Biofuel is also a source of renewable energy and the one most people know the least about. This concept involves using types of materials that you can burn to create energy. This can be left over paper and wood, trash, and even manure from animals.

It is quite an interesting concept and one you will want to be sure you read about. You can be sure this is one area of renewable energy that will continue to grow by leaps and bounds in the next decade. Don't underestimate how valuable it can be as it also removes waste from our environment. This type of renewable energy is scientific in nature and one that has been around the least amount of time.

Option 1: BioFuel

Biofuel isn't one of the well known types of renewable energy but an important one to understand. The process begins when plants to through photosynthesis. There is chemical energy stored inside of it that can be released. What they create is a type of biomass and that can be turned into fuel.

It can then be burned in combustible engines. There is still plenty of research that needs to be done in the area of biofuel. The process right now of converting it isn't as effective as it should be. It is also extremely time consuming and expensive to do so.

Biofuel can be in the form of a liquid or a solid. Vegetable oil that is used as an alternative fuel source for some vehicles out there is a type of liquid biofuel. It can be natural or it can be reprocessed after it has been used. Some restaurants give their used vegetable oil to those that burn Biodiesel in their vehicles. Once it has been cleaned they are able to use it without harming their vehicle.

Some types of food items are grown in higher supply than the demand just so that ethanol can be produced. It is usually mixed with about 15% regular gasoline in order to make the mix work. It seems that many in the business of growing such food items though don't always agree with this use of it. They feel the foods should be used to feed those in need.

Biofuel is most commonly found in the form of a solid though. For example burning wood falls into this category. You can use it to cook with and to heat your home instead of relying upon natural gas. The downside though is that this can emit dangerous elements into the environment.

There is ongoing testing where the variables are being controlled right now in the area of biofuel. It is believed this type of renewable energy could one day be a breakthrough in the area of supplying fuel for our vehicles. Look for great things to be coming up in this area in the future.

If we can figure out an affordable process for using biofuel we can create more than half of what is depleted annually from our natural resources. This is what facts continue to motivate researchers to move forward and funders to continue with grant money to allow it to happen.

The one issue that seems to be a concern is that in order to generate more biofuel to use, there is a great deal of land that has to be accessed. That could mean land normally used for growing food and other resources is no longer available. A close eye will need to be kept in that particular area.

As some of this concern has come to light, other forms of biofuel to be able to use are being introduced. Since these natural products are able to give off heat that can be transferred to fuel. Even trash can be used to create biofuel. This means less of it will be around to remain in landfills.

One of the most successful biofuel companies out there is in Cedar Rapids, Iowa. It is known as BFC Gas & Electric. They are able to recycle approximately 150 tons of materials each day at their facility. They process wood remains from projects and from sawmills in the area.

There is also paper that is used instead of being wasted when it was used for projects and scraps remain. Sometimes there are types of paper and cardboard that can't be successfully recycled so it is processed here.

Crops that have been ruined, diseased plants and trees, and the corn stalks that remain after harvest all work as well. More than 40,000 homes in the Cedar Rapids area are provided with electricity from this particular company. They are really working hard to put the use of biofuel to work for the benefit for their community.

In some other areas the use of changing cow manure into biofuel is being done. This is something that not everyone finds appealing but it can be a viable way to get more benefits out of such waste. There are still many details of this type of biofuel to be worked out though.

Option 2: Hydropower

There is also power found in water and that process is referred to as hydropower. It can create much more energy at one time than either solar power or wind power. This is due to the fact that water is so dense.

Therefore it only takes a little bit of water to create some power. It takes much more sunlight or wind to create that same amount. Hydropower is used to offer electricity to more than 28 million people around the world. It accounts for approximately 10% of all the electricity produced in the United States.

The process involves plenty of safety precautions as the water can never come into contact with the electrical part of the process. We all know that can spell disaster. The process occurs as the water flows and it spins turbines that are found in a generator.

While this process isn't new, it has been upgraded since it was first used thousands of years ago. The Egyptians were a very smart society and they were able to use it to grind down their grains. It was also used as a way to saw wood as it kept the blades cool enough.

The first hydro power plant in the United States was introduced in 1882. It was placed in Wisconsin. Most people don't realize that a great deal of the early electricity and power in the United States once came from such hydropower plants. In 1940 approximately half of the electricity used came from this source. Later it was replaced by the use of coal.

Of course that was before we knew that coal was something that we will one day run out of. We didn't realize at that time just how much of it we were going to be using. We also didn't realize how much it was going to pollute our environment. It seems ironic that to move forward we have to go back but that is how it plays out in this particular case.

There are several dams in the United States where elaborate designs are in place. Here huge volumes of power are generated each day using these methods. It is a very time consuming and complex process though to turn water into energy. Since hydropower is so expensive it isn't as widely used as it could be.

The largest hydropower plant in the world is located in Washington along the Grand Coulee Dam. It took 11 years to build it from 1933 to 1942. Since that time it has undergone numerous facelifts and changes. It is amazing at a size of over 5,000 feet long and 550 feet high.

In all there are four distinct power plants found here. Each one is controlled separately for easier control of the process to make energy. There are 33 generators and more that 2 million homes in that area get their energy from this particular location. It is an amazing accomplishment and one that really shows the power that can be created with such a renewable energy source.

Some of the newer approaches for hydropower include generating that power from the waves in the ocean. Even the waves that occur all day long in the waters there can be used to create energy. It is believed that the use of hydropower will one day be much more significant than it is right now.

Of course it doesn't make sense for hydropower to be used everywhere. There are many places where there just isn't enough water flowing in the area for them to work well. Not every place as an abundance of rivers, streams, and oceans to produce hydropower. Most of them are found along the West of the United States. They are in Oregon, Washington, and California.

If a way to convert that energy from the water can be found that is less expensive than what we know right now it surely will. It only makes sense due to the volumes of energy that can be created from this particular natural resource. It also doesn't reduce the amount of water out there or place any type of harmful contaminants into it.

There is research though that indicates the aquatic live forms in the waters may be affected by the process. It can upset the natural balance of what ordinarily goes on in that water. There is also the risk of terrorism in an event to destroy what we have built.

Since the events of 09/11 tighter security efforts have been implemented in order to prevent that from happening. There are random checkpoints and stops along the way where vehicles are inspected. There are also restrictions at places including Hoover Dam that prevent large trucks and buses from going across them as an additional safety precaution.

It can take years to build the dams for a hydroelectric plant to start operating though. Many people who live in the vicinity of a proposed hydroelectric plant aren't happy about it either. They worry about flooding and they often find their homeowner's insurance goes up due to that possibility.

Option 3: Solar Power

Solar power is likely to be the most valuable of the renewable energy sources available. That is because it is plentiful and it is the least expensive to implement. New buildings can be built that incorporate the concepts as can homes. There are also solar panels that can be added to what is already in place.

That way you can take any home or business and modify it to use solar power with. They can be expensive but you will find that they can be a very good investment. If you own an apartment building with the electricity included in the rent this can help to reduce your overhead expenses. In a couple of years the process will pay for itself.

This process involves placing collectors and panels that will be in places where the sun will reach such as the roof. The process of converting what is collected into energy is known as photovoltaics. In order to make this happen, silicon is used to transform it.

The concept of solar power has also found its way to vehicles. They have cells that capture energy from the sun to provide movement for the vehicle. The advanced design allows that energy to be continually conserved too such as when you come to a stoplight.

If you drive your vehicle for longer than you have enough solar power for you aren't going to be stranded in the middle of the interstate. You also won't find your vehicle shutting off the moment the sun goes down. This is because it also operates on regular gasoline.

The solar energy is always consumed first though before any of the gas is used. You will continue to use the gas for movement until you have replenished the solar energy. Even when you are using the gas though this type of vehicle is going to burn cleaner and thus not place all of the pollutants into the atmosphere that other vehicles do.

Hybrid vehicles are available in quite a few models and more of them seem to be introduced all the time. However, it can be very expensive to be able to purchase one. They cost quite a bit more than other types of vehicles on the market. Yet if you are looking for a great way to rely upon renewable energy sources this is one way to do it.

If you are paying a fortune right now for fuel to keep your current vehicle going it may be most cost effective for you. Even with a higher monthly payment until you pay off the hybrid vehicle you can take the money you save on gas and allocate it towards that payment.

Solar power plants have started to crop up in locations where people never expected them. Both California and Florida have plans to create at least eight new locations each in the next five years in order to generate more of their power from the sunlight.

In fact, California is a leader in promoting the use of solar power. There is a plan in the works to create 3,000 megawatts of it by 2017. The goal of this is to create a way for California to get all the energy they need without using up their natural resources.

Many of the citizens of California aren't happy with the initiative though due to the cost of it. This is a project that will cost almost $3 million before it is completed. Yet the leaders of California are confident it is a step in the right direction. They are confident it will help all of their residents now as well as future generations.

Many people in California though are taking the initiative and running with it. They are being offered very high rates of money back for installing solar panels on their homes and businesses. Construction companies are offered incentives to build new structures with them already in place.

California isn't the only state that is moving forward with incentives for residents though. Check to see if your state is willing to reimburse you for a portion of the cost of having the solar panels installed. This can be a great way for you to save money on the project. At the same time you will be doing your part to cut back on the amount of non-renewable energy your household consumes.

It is vital that we learn do to all we can with collecting solar power. Experts believe that there is more energy from the sun every single day that we can use than what we burn up in fossil fuels annually. That is an amazing evaluation and one that does mean a great deal for the future of all of us.

Just imagine if we were collecting and using a fraction of that amount what we would be able to preserve in regards to our non-renewable resources. You may be asking just what is holding everyone back and the answer is simply the cost that is involved.

Yet as you will notice more and more of it will be taking place and every little bit will be one step closer to really getting the results we are striving for as a society. It is going to take decades for us to get there but that is still better than the alternative. There is no reason to continue depending only on non-renewable energy when we don't' have to.

It is estimated that 20,000 homeowner's in the United States currently get at least 80% of their electricity from solar energy. This saves approximately 50,000 tons of coal each year in order to produce that same amount of power. That is quite a bit of our natural resources being saved from such a small number of people being involved.

Pros & Cons Of An Isis Free Solar System

Isis solar limited is offering free solar system installation and maintenance on residences. While on the surface this sounds like an extremely good deal for people wanting to save on their energy costs like everything else it has its upside and downside. This article will explain a little bit about how the free electric system works and the pro and cons of having a free system installed.

How the Free Solar System Installation and Maintenance Works

If your home meets certain criteria to have a free solar system installed Isis will do so and maintain it for the next 25 years. Here are the ground rules.

- You must have an un-shaded roof facing South with enough space for the installation of the system.
- You must agree to keep the system for 25 years and to allow access for maintenance of the panel.

In return for installing the solar system on your roof and maintaining it for free Isis gets the profits from all the excess energy your solar panels produces that is not used by you. These profits can be quite large over the next 25 years. However, if you cannot afford to have solar panels installed or do not have the credit to do so, the savings in your own energy bill may make the installation of the free panels and your energy savings worth it.

The Pros of Isis Free Solar Systems

- The main pro is that you will save money on your own energy bill. How much you will save will in part depend on how much energy you use during the day time hours. If you are home all day running the vacuum, the clothes washer and dryer, a dish washer and uses a large amount of electricity you can experience a big drop in your electric bill. 30% or more.

- For those of you who really want to make a difference in the environment. These free solar systems will not only provide you with a green way to have electricity but, you will also be helping supply electricity from renewable energy to others as well. Lowering the overall carbon foot print.
- At the end of 25 years the solar system is yours to keep and you may be able to not only continue saving on your energy bill but, may be able to make a small profit from the excess energy your system generates.

The Cons of Isis Free Solar Systems

- You lose all potential profits generated from the solar panels placed on your roof. Isis keeps those profits and does not share them with individual home owners. If you can afford to pay to have the same system installed experts predict that between the savings on your energy and profits you can make from your excess energy you can be completely reimbursed for your panels in 9 years and start making profits from the excess energy sooner.
- In some cases homeowners with solar panels find selling their home more difficult which means that having Isis install a free solar system that you are required to keep for 25 years may mean that you won't be able to sell your home if you want or need to?

Before signing on the dotted line for a free solar system weigh the pros and cons. While these free systems can be a great deal they are not right for everyone so make sure they are right for you.

Option 4: Wind Power

Across the plains where there is open land, the wind is able to move at high speeds. This is due to there not being any buildings or homes in the area to break it up. This is the perfect location for wind power to be generated and collected.

The use of windmills to create energy has been around for a very long time. Of course these windmills were quite small compared to what is out there just to create the energy. Not all spaces are right for large windmills so they look for those where the wind constantly blows at a rate of at least 15 miles per hour.

There are large windmills hundreds of feet high strategically gathered in these open areas. Each of the blades on the windmills requires a semi truck to haul it. They are then put into place with various pieces of construction equipment. They operate on their own but are closely monitored to ensure everything is working like it should on them.

Even though it is a great deal of cost and work to get them installed, each one can last about 20-25 years. That means they are well worth the investment because they don't constantly have to be replaced. That really has been an encouragement for more of them to be put into use.

The collection of wind power has proven to be so profitable in many small towns that some farmers and ranchers have given up those operations. Instead they are allowing various companies to place these windmills on their land and getting paid to do so. They don't have to do any of the work to maintain them and they are guaranteed to make money.

With the high risk involved in the line of farming and ranching, it is understandable why so many of them do it. For those close to retirement age as well it just makes sense to be able to rely on a given income source. This is less stressful than hoping for a good crop in the field or that the price of meat doesn't decline on the market.

Installing these types of windmills to generate the power though is extremely expensive. It is also a time consuming process due to the sheer size of them. When you see them from an airplane or driving down a main highway you won't realize just how enormous they are. On average they are approximately 200 to 250 feet from the ground.

The largest wind turbine in the world is located in Hawaii. It is more than 20 stories high. Each of the blades on the windmills are the size of a full football field. Each of the turbines located here is able to produce enough energy to offer electricity for 300 homes in the area.

If you get a chance though go drive into an area that features them so you can see with your own eyes how they really appear. In Iowa there are approximately 800 of them on what is know as wind farms. More than 200,000 homes are supplied with power from them. You will also find them in Minnesota, Wisconsin, Oklahoma, Kansas, California, and Colorado.

When you watch the turbines spinning you will notice the turn counterclockwise. As the wind blows the shaft is engaged which is connected to a generator. As the blades turn this generator will move the power created along to the locations on the grid. They are monitored and there is an emergency shut off for them as well.

In the event of pending natural disasters such as a tornado, these turbines can give them more power to move on and to destroy everything in its path. By shutting them off that isn't going to be the case.

It can also prevent the blades from being damaged during such an event. With the cost of the windmills being so high, keeping them in good working condition is very important. In some areas tornadoes are a common problem so the windmills are constantly watched.

Not everyone is a fan of these windmills though because they think they are an eyesore. They want to be able to gaze out their window and see the open land. Instead they are seeing tons of windmills. They also feel it takes away the romantic and tranquil feeling of the area.

Yet when you have small towns that are struggling for their residents to make a living, the prospect of getting paid to let a company come in and place windmills in place to collect the energy is an answer to a pray. They would rather have money for food on the table than to have the wide open spaces.

The amount of energy that can be produced this way will never be the same. It can be severely windy one day and only a gentle breeze the next. Even so, there is enough of it to make it worth the cost of getting all the equipment in place. It is worth it to be sharing the natural beauty of an area as well.

Some of these turbines are also located in various bodies of water. It is believed that wind power could one day be responsible for about 50% of the energy we consume. There is still a great deal of research and development that must take place for it to happen.

Yet it is very exciting to know such a renewable energy resource can be there for us. It can make us breath easier when we continue how much of our natural resources that we can't replace continue to be consumed each and every day.

The fact that it doesn't release anything negative into the environment is very important as well. The fact that here isn't carbon dioxide and other gases going into the air as wind power is produced should mean a great deal to each of us. It means the air we breathe is healthier for us than it was before.

The American Wind Energy Association has continued to work to get this type of renewable energy in place around the country. Many find it strange that other countries including Spain and Germany rely on it more than the United States does. However, with the positive aspects of it you can expect the volume of it used here to continue increasing annually.

According to their estimates, we can create 1 ½ times the amount of electricity used annually in the United States based on the amount of wind that is out there daily on average. That is very exciting news and something that could really help us to preserve our natural resources to the best of our ability.

There have been some animal rights groups pointing out that the spinning turbines on these windmills is seriously injuring and even killing various types of birds. To evaluate these claims there have been some studies in place.

Efforts are being made to prevent this without reducing the effectiveness of the wind energy that is being produced. It is believed that these incidents are very limited though and that birds aren't in any real danger due to the turbines out there creating energy.

Our Future With Renewable Energy?

As you can see there will be plenty of benefits from all of these types of renewable energy. While they are all implemented right now on some level, there is still more that needs to be done. We need to learn how to get the maximum benefits from them.

At the same time we need to learn how to reduce the costs involved with getting that energy out of them. That seems to be the biggest hold up with getting them really rolling. The fact that they work and they reduce pollution is very positive aspects that encourage us to move forward with them.

As we come more aware of the fact that we are depleting our energy sources we need to take action now. The more we can learn not to depend on those resources the better off our entire world will be. Take your time to learn all you can about renewable energy too so you can be aware of what is going on around you.

Keep up on the developments that are taking place in these areas. It is actually quite fascinating and certainly something you will want to continue exploring. At the same time it is wise to take a close look at how you are using natural sources of energy.

Make cut backs where you can to continue conserving what we do have available right now to create the energy we often take for granted in our society. It doesn't mean you can't continue to enjoy your lifestyle, but rather try to view it from an environmentally friendly perspective.

There is some negativity out there though in regards to producing renewable energy. There is a concern about the amount of space it is going to take to get all of these methods into place. There is also the concern that some businesses will lose money as they won't be needed in full demand to harvest the fossil fuels. Yet the overall design of the turbines for windmills and for solar panels is an area where improvements can continue to be made.

If they don't have to take up as much room or be so noticeable more people will be willing to install them. Getting creative in this area is something we should encourage and that should be possible due to the technology we have readily available to us.

Some of the experts also worry that there will be some problems that arise but we aren't aware of them yet. However, that shouldn't prevent us from moving forward and getting all we can from these renewable energy sources. Crossing those bridges as they arise is the best course of action to take.

They are looking though along the lines of pollution and destruction due to the use of the types of construction equipment used to complete such projects. There is also a concern about the risk of the work too. Even with proper training people can get hurt or killed in the process of erecting them.

Some people aren't up for change and that includes how they get their energy. They are used to relying upon coal and natural gas. They are used to what they pay for these items and they are afraid to embrace something new. Continuing to provide them with accurate information though can really help them to understand the big picture that is here.

The possibilities that are open to us along the lines of renewable energy are huge. There is no limit to how far we can take them as long as they are going to benefit society as a whole. It will be very interesting to look back ten or twenty years from now and see how far the concepts of renewable energy have come.

ps You Can Apply Today

1. It begins with your commitment.

If you want to save energy, you have to be committed enough in doing the necessary things for it. Without your commitment, you may not become mindful of you actions, which can directly or indirectly affect your overall energy consumption. Thus, you have to be committed, so that you can see results on it.

2. Make it a practice to turn off the lights in your room.

Lots of people today leave their bedrooms without turning the lights off. Although lights do not consume much electricity, lots of energy would be wasted if you constantly left them turned on for the most parts of the day, even if you are not inside the room. Thus, make it a practice to turn off the lights, so that you can save energy.

3. Replace your filters regularly.

Most heating and cooling equipment require clean filters, in order for them to run clean. The filters can also help them in becoming more energy efficient. With that, replacing your filters regularly can help you save a lot in terms of energy consumption. Ask your technician about it, so that you would know when is the best time to do so.

4. Dirty coils make your appliances consume more electricity.

Refrigerator coils can get dirty over a certain period of time. In most cases, they can accumulate a lot of dirt within the six month period; and, when that happens, the dirty coil would make your refrigerator work harder in order to achieve its desired temperature. Thus, it is best if you practice cleaning and vacuuming your refrigerator coils every six months or so, so that you won't have to endure higher electricity bills.

5. Close the windows.

Whether you need to achieve lower or higher temperature inside your house, it is best if you close your windows. This is because, open windows would make your cooling or heating equipment require more energy to serve you better. When you close the windows, your equipment would not work too hard, which means it can help you save energy.

6. Do not forget the timer.

When you sleep at night and you are using your air-conditioning system to make your room temperature cooler, it is a good idea to make use of its timer. The timer would ensure that your AC would be turned off in time. Aside from that, if you won't use the timer, you may get tempted to extend its usage, especially when you don't feel like going out of your bed yet.

7. Make use of solar energy.

There are lots of DIY solar panels that you can make use for your home today. One of the best things about this new technology is that, you no longer have to pay high electricity bills with them. All you have to do is to have someone install it properly, and you should be good to go.

8. Dry your clothes the natural way.

One of the best ways to conserve energy is to dry your clothes naturally. If you are going to use your washing machine's dryer for them, then you would be consuming lots of energy. Just hang the clothes outside, and let the sun dry it so that you can save more energy.

9. Unplug all unused appliances and electronic devices.

You need to keep in mind that many appliances and other electronic devices consume electricity when they are plugged into the socket, even when they are turned off. Thus, the moment that you turn them off, you should ensure that they are unplugged. If it is a hassle for you to do, then just do it just before going to bed.

10. Your computer monitor.

If your computer's monitor is still the large type, or the CRT, then it is time to replace it. CRT screens actually consume more energy than the LCD types. Thus, it is time to make use of the newer computer screens.

Aside from consuming lesser amounts of energy with them, it also improves your whole experience in using your desktop PC.

11. Open up your windows during summer.

Summer heat can be hard to beat, which is why it is best to come up with ways to get around it without consuming lots of energy. Open up your windows during summer time, so that you can let fresh air go in and out of the house. This is effective, especially if your house is surrounded with trees.

12. In going for a vacation.

When you go for a vacation, you need to make sure that all the appliances, except the refrigerator, are turned off and unplugged. Aside from that, if you have a water heater, then you should also ensure that it is turned off as well. Minimize energy consumption by leaving just a light in the porch and in your living room while you are away.

13. Minimize the number of times you open the door.

When it is warm and you are using your air-conditioning to achieve cooler temperature inside your place, then you should minimize the times that you and other members of your household are opening the door. Each time you open the door, warm air would get inside your place. When that happens, it would make your cooling equipment work harder.

14. Turn off your light when you go to bed.

Turning off the light inside your bedroom when you go to bed can help save lots of energy. If you are not very comfortable, you can always use a lampshade instead or a night light. When you do this, aside from saving energy, it can also help you sleep better.

15. Using water heater.

If you have a water heater installed at your place, it can make your energy consumption go up, if you have leaky faucets. Thus, you should fix any leaky faucet as soon as you spot them. Aside from saving energy, it can also help you save on water usage.

16. Cut off cooling areas in your house that don't need it.

When you turn on your cooling equipment at your place, you have to consider the fact that it would try to lower down the temperature on the entire space. Thus, if you have rooms that don't need cooling, then you should close the door to that room. By minimizing the space that it has to cool down, you are ensuring that it would not consume more energy that it needs to.

17. Block the summer light to save energy.

Blocking the summer light can go a long way, as far as saving energy is concerned. This is because, it can help in bringing down the temperature inside your house. Thus, you need to install awnings, shades, blinds, or sunscreens in spots where sunlight can enter. Reflect away the heat from your house to make it cooler.

18. Have a technician check your electric wirings on a regular manner.

Checking your electric wirings inside your house is necessary, in order to lessen energy consumption. Routine check up would ensure that no wires are damaged by pests, which can cause more energy consumed or even short circuit. Thus, this practice would not just conserve energy, but also make your home safer as well.

19. Installing an exhaust fan in your kitchen.

Whenever you are cooking, an exhaust fan inside your kitchen can help a lot in blowing hot air from the place. This can help you save a lot of energy especially during summer time. Your savings when it comes to using energy to cool down your place would be more than the electricity consumed by the fan.

20. Air dry your dishes whenever you can.

Using the dishwater heater consumes a lot of energy. Thus, it is best if you turn it off, whenever you leave your place and also when you are not making use of it. Aside from that, it is also a good idea to simply air dry your dishes whenever you can or when you have the time for it.

21. Be mindful of your cooling system's thermostat levels.

It is best to be mindful of your cooling system's temperature or thermostat levels. This is because, it can determine the amount of energy that it would consume in cooling down your place. Thus, you should take note of the comfortable temperature that you want to achieve, so that you won't go lower and consume more energy in the process.

22. An alternative to using your air conditioning system.

If it is warm and you don't want to turn on your air-conditioning system for a while, you can make use of an alternative for it. All you have to do is to place a bowl that is filled with ice cubes in front of your electric fan. Turn on the fan, and you can feel cool air blown towards you.

23. Reduce the humidity inside your place.

Reducing the humidity inside your place can greatly help in making you more comfortable during warm weather conditions. To achieve that, all you have to do is to use a dehumidifier. When you make use of it, you can minimize the number of times that you have to turn on your air-conditioning system.

24. Use your air-conditioning system's timer.

It is a good idea to make use of your air-conditioning system's timer on a regular basis to save energy. For example, when you turn it on at night before going to bed, set the timer to turn off the system by dawn. This is because, it is usually pretty cold during that time, and you won't even notice that your AC has already been turned off.

25. Minimize the use of your shower's heater.

There are times when people make use of their showers' heater even on summer times. Thus, whenever it is warm, you should see to it that your shower's heater has been turned off. Do this, so that you can enjoy cooler water, and save on energy consumption.

26. Minimize the times your kids open up the fridge.

Whenever you open the door of your refrigerator, its cool air would go out and warm air would go in. Therefore, it can make your fridge work harder, and consume more energy because of it. Thus, you have to remind your kids not to open up the fridge a lot. Aside from that, let other members of your household be aware of it, so that they can all help out.

27. Purchase an air-conditioning system that can help you conserve energy.

There are air-conditioning systems today, which are designed to help you conserve energy. One example of it is that, it has a plug that comes with a socket in it, in which you can plug your fan into. With that, once the timer turns off the system, your fan would automatically be turned on, so that you won't even have to get up of your bed to do it.

28. Buy more food items to place inside your refrigerator.

Keep in mind that a cooling system usually works harder with larger space. Thus, if your refrigerator is half empty, then it needs to consume more energy to cool it down. With that, you should try to consume more space inside your refrigerator, so that you would be able to save energy.

29. Turn off your computer.

Lots of people think that when they turn their computer off and on, it would wear out the system faster. However, it is actually untrue, especially with the newer computers. Thus, it is time to make it a practice to turn off your computer whenever you are not using it, so that you can bring down your electricity bills.

30. Turn off all heat sources in summer.

One of the best ways to minimize heat during the summertime is to turn off all the heat sources inside your place. Some of which would be lights, appliances, and other electrical items. Turning off these things whenever they are not in use would minimize heat inside the house, and would help you save on energy cost.

31. Use LED for your lighting needs.

LED lighting is now being used by a lot of companies, in order for them to reduce their electricity bills. This is because, these types of lighting are energy efficient compared to traditional types. Thus, it is time to make use of such types of lighting, so that you too can enjoy more savings from them.

32. Add light timers in your place.

Reducing the amount of time you make use of the lights inside your place can help a lot when it comes to energy saving. Thus, installing light timers can help you achieve your goal in it. With a timer, you can ensure that the lights in your porch or garage would be turned off, even before you go out of bed.

33. Energy saving in cooking.

Whenever you are trying to cook something with the use of a pot, it is better to make use of the lid for it to conserve energy. By using the lid, it ensures that heat would not go out of the pot and be focused on the food that you are trying to cook. This is actually one of the reasons why the lid is always included whenever you purchase a pot.

34. Having a swimming pool.

A swimming pool can greatly help you during summertime, since it can reduce the amount of time you make use of your air-conditioning system. However, running your pool can also consume lots of energy, especially if you let the water pump run 24 hours a day. To get around it, just let it run for 6 to 8 hours each day instead, so that you can save energy.

35. Use compact fluorescent light bulbs.

Using the traditional incandescent light bulbs use a lot of energy. This is because, aside from providing light, they also produce heat. Thus, it is time to replace them with compact fluorescent light bulbs today, so that you can save more energy. They do not produce heat, and they consume 80 percent less energy.

36. Motion sensing devices.

Motion sensors can help you a lot when it comes to conserving energy. This is because, you can make use of them for your outdoor lighting. With motion sensors, you can ensure that your outdoor lighting would only be used whenever the sensors detect something moving outside your place.

37. Your furnace filters.

Keep in mind that your furnace filters can get dirty in just a matter of time. The dirtier they get, the more energy your furnace fan needs to consume in order to work. Thus, it is best if you replace the furnace filters on a regular basis, so that you can consume less energy.

38. Give your computer a break.

Keep in mind that your computer becomes slower as you fill it up with more and more data. Thus, it is time to clean out unwanted data from it. By doing that, your computer would work faster, which would reduce the time you need to spend in doing something in and it that would result to more energy saved.

39. Get your family to watch television in the living room.

It is quite a usual thing these days for each of the members of the family to have their own television set insider their own rooms. Whenever you are all trying to watch the same TV program though, convince all the members of the family to watch it in the living room. When you do this, you won't just be saving energy, but it can also give you time to bond with each other.

40. Buy energy efficient appliances.

If you are shopping for appliances, you would realize that in most cases, the energy efficient products are usually more expensive. Although they are, they can actually provide you with more benefits in the long run. Aside from helping you minimize energy consumption, these models usually are also made in higher quality than the others.

41. Charge your mobile phone's battery, when it is almost empty.

Lots of people today put their mobile phones with its charger every time they arrive at their place, even when the battery is still half full. It is actually a better practice to charge your battery, when it is almost empty, so that you can minimize the number of times that you plug it into the socket. Aside from that, it can also prolong the life of your battery.

42. Using the right amount of detergent.

When you make use of your dishwasher, you should see to it that you are putting just the right amount of detergent. This is because, too much of it or too little can affect your dishwasher's efficiency. It is time to read the instructions well, so that you can conserve more energy.

43. Use a dishpan whenever you are hand washing dishes.

Hand washing dishes is actually a good way to save energy, especially if there are only few of them that you need to deal with. However, you should not let hot water run continuously whenever you are doing it. To avoid this, you can always make use of a dishpan, in order to hold water. This would not just minimize the energy consumed, but it would also conserve water.

44. Using an electric oven for cooking.

When you are using your electric oven to cook something up, you can always turn it off in advance. In other words, you can turn off the oven even when you are not done cooking yet. There is no need to worry in doing that, since the oven can retain heat. This can also be done when you are just trying to heat your food.

46. Do not over dry your clothes.

Whenever you make use of your washing machine's dryer, you have to be careful not to over dry your clothes. This is because, over drying them would waste a lot of energy. Aside from that, it can also ruin your clothes if you do it all the time.

47. Cook barbeque and steaks every weekend.

To enjoy your weekends, you can always cook barbeque and steaks with your family. This would reduce the amount of time that you make use of your electric stove and oven. Aside from that, it would also provide you a chance to spend time with your spouse and kids.

48. Have smaller cooking appliances.

Unless you are constantly throwing a party or there are lots of members in your family, you should make use of smaller cooking appliances to save energy. This is because, smaller ones simply consume lesser energy to do the same cooking task. Thus, it is time to make use of the smaller appliances, and use the larger ones only when you really need them.

49. Moving your refrigerator.

Always remember that the temperature around the place where your refrigerator is situated, can affect the appliance's energy consumption. Thus, it is important that you situate your fridge in a cooler place. Do not let it stand near a cooker or where the sunlight comes in, so that it won't have to work harder to cool down the food items it contains.

50. Do not forget to defrost your refrigerator regularly.

It is important to defrost your refrigerator on a regular basis. This is because, it can greatly help when it comes to its efficiency. Aside from that, it can also help in prolonging its life. If you don't want to defrost manually though, then it is best to buy a refrigerator that has an automatic defrosting feature.

51. Taking a shower instead of a bath.

According to a lot of experts, taking a shower actually conserves more energy than taking a bath. To ensure that this is the case though, it is best to make use of low-flow showerheads, instead of the power showers. They provide the same comfort and consume lesser amounts of energy.

52. Make it a practice to avoid using hot water.

Heating water takes a lot of energy, which is why it is best to make it a practice to avoid using it. Lots of people heat water to take a bath. However, if you make it a practice to use regular water, then you can actually get accustomed to it eventually. Thus, if you want to save energy, then you should avoid using hot water from now on.

53. Insulating your home.

Insulating the ceiling and walls in your home can actually help you save about 25 percent of your heating cost. However, it needs to be done properly. Thus, you have to research about it, if you want to do it yourself. On the other hand, you can also hire someone to do it for you.

54. Walk.

Keep in mind that using your car or motorcycle also requires energy. Thus, if you are trying to get to a place that is just walking distance from your place, then you should avoid using them. Aside from conserving energy, it can also help in saving the environment by using lesser amounts of fossil fuel.

55. Buy recycled paper products.

There are lots of recycled paper products available in the market today. When you purchase them, you are basically saving energy, since it takes about 70 to 90 percent less energy, when it comes to recycling paper. Aside from that, you are also helping in preventing the loss of trees in a worldwide scale.

56. Do not throw away your shopping bag.

Making use of reusable bags when you are shopping can actually help in conserving energy. Aside from that, it can also help in reducing the amount of waste that you introduce to the environment. Thus, it is time to make use of reusable bags, instead of accepting the disposable kinds each time you shop.

57. Plant a tree in your backyard.

When you plant a tree in your backyard, it would eventually provide you with shade, which can reduce your air-conditioning bills during the summer time. Aside from that, it can also minimize the amount of carbon dioxide in the air. Thus, it does not only help you reduce your bills but also help the environment.

58. Use renewable sources of energy.

When you make use of renewable sources of energy for your needs, you would be able to save more money when it comes to your electricity bills. Some of these sources are the wind and solar. This move can also reduce the amount of fossil fuel burned, which is helpful to the environment.

59. Start a carpool.

When you share a ride with your coworkers or friends, it would actually reduce the amount of fuel that you would consume. There is no need to do it every day at the start. You can do it at 2 days in a week at first, and once you get the hang of it, increase the number of times that you do it, so that you can all help in saving energy.

60. Arrange your furniture items properly.

It is important that the air circulation from your cooling or heating equipment is not being blocked by anything. This is to ensure that it works at its maximum efficiency. Thus, you have to arrange your furniture items with that in mind, so that you can ensure lesser energy consumption.

61. Use your fireplace.

During the colder months, do not forget to make good use of your fireplace. By doing that, you won't have to make use of a lot of energy in order to bring up the temperature inside your place. Just take out the fire though once you go to bed, so that you can also conserve firewood.

62. Kitchen fans.

Whenever it is hot and you have turned on your air-conditioning system, you should not make use of the electric fan that you have in your kitchen. This is because, it can make warm and moist air go towards different parts of your place, and make your AC system work harder.

63. Clean your air-conditioning systems regularly.

Cleaning the filters of your air-conditioning system and cleaning the system itself are two different things. When it comes to cleaning the AC, it is best if you hire someone to do it for you. Do it regularly, so that it would be able to work as efficiently as possible.

64. Use your pressure cooker.

If you don't have a pressure cooker, then it is time to purchase one. This is because, using a pressure cooker can help you save energy. With a pressure cooker, you would be able to cook food faster, since you would be applying pressure on it, aside from the fact that it would also make the temperature go up fast.

65. Inspect the indoor and outdoor coils of your air conditioner.

There are two coils that you need to inspect regularly when it comes to caring for your air conditioner. The indoor coil should be checked and cleaned always, since it gets wet during the process of cooling, aside from attracting dust. Dirt buildup on it can make the AC work harder, which is why it should be kept cleaned always. This is also the same with the outdoor coil.

66. Prevent excessive cooling.

There are times when you set the thermostat of your air conditioner beyond your desired temperature. You may simply become adjusted to it by using a blanket when you sleep. However, it is best if you bring down the thermostat level instead, since making your room cooler than you need it to be would make you consume more energy.

67. There is no need to make use of air conditioners during the entire summer season.

Although it is summer, there are times when the weather can be milder. It is during these times when you can substitute your air conditioner with just an electric fan. By doing that, you would be able to reduce your energy consumption by 40 to 60 percent.

68. Choose the right power for your room air conditioner.

Room air conditioners are available in different horse powers these days. You can choose from half, one, one and one half, and many more. When it comes to this, you need to consider the size of your room to be able to select the right horse power for your air conditioner. It is best that your AC has just the right power for your room, since too low or too high can make you consume more energy.

69. Properly installing your air conditioner.

It is very important to ensure that professionals are the ones to install your air conditioner. This is because, its installation is a factor that can affect its efficiency. The air conditioner has to be installed on a flat surface, since it would help its drainage system and other mechanisms to work properly. See that it is installed properly, so that you won't be consuming more energy than you should be.

70. Using a window air conditioner.

If you are using a window air conditioner, it is best if you set the fan speed on higher levels. However, if you can feel high humidity, then you should set it to low. Lower fan speed can actually make you feel more comfortable during warmer months, since it can help removing moisture from the air.

71. Check your car's tires weekly.

You should make sure that your car is fuel efficient, in order to save energy. One of the factors that can affect that is its tires. With that, you need to check the tires of your car weekly, so that you can ensure that it has proper inflation. When that is the case, then you would be able to save energy by making your car more fuel efficient.

72. Read more books.

Reading more books can help you reduce your energy consumption. This is because, it would greatly reduce your time spent in front of your television. With that, it is best if you encourage your kids and your spouse to follow your lead, so that they can all contribute in saving energy.

73. Do not place your TV set or your lamp near your air conditioning thermostat.

It is not a good idea to situate a television set or a lamp near your air conditioning thermostat. This is because, these appliances can give off heat towards the thermostat. When that happens, the thermostat would try to work harder than necessary to cool down your room's temperature.

74. Make your company as paper less as possible.

Making your company as paper less as possible can help you conserve energy. This is because, each time you need to print something, you would be consuming more electricity. Aside from making your company reduce electricity bills by becoming paper less, you can also help the environment.

75. Unplug your air conditioner during winter.

It is a fact that you won't be using your air conditioner during the winter times. Thus, it is best if you can unplug it, so that you can ensure that it won't be consuming any kind of energy. It is also better if you can cover it with something, so that you would be able to preserve its quality.

76. Do not forget the seal between your AC and your window.

It is important to constantly check on the seal between your air conditioner and the frame of your window. This is because, the seal can get easily damaged with moisture. When that happens, the seal can eventually have holes, which would let cool air from your room escape.

77. Drain water from your hot water tank.

Draining water from the hot water tank regularly can help a lot when it comes to saving energy. This is because, sediments inside the tank can reduce the efficiency of your tank. Draining even just a quart of water from the tank once every two to three months would already help a lot.

78. Cover your foods when you store them inside the refrigerator.

It is important to cover or wrap foods that you store inside your refrigerator. This is because, foods that are uncovered can release moisture inside the cooling unit. When that happens, it actually makes the compressor work harder, which means more energy consumed.

79. See if you have a butter conditioner insider your refrigerator.

It is always best to check on your user's manual, even when it comes to making use of your refrigerator. This is because, it can also help you determine if your unit has a butter conditioner. This part of a refrigerator is actually a heater, and it is best if you can turn it off.

80. Don't keep your old refrigerator running.

It is possible that you are still using your old refrigerator at your garage, especially in times when you need extra space. However, it would actually serve you better if you just obtain a new unit, even just a small one. This is because, old refrigerators can actually consume a lot of energy, due to their being inefficient.

81. Attain the ideal temperature for your refrigerator.

If you are going to make your refrigerator cooler than you need it to be, then you would be consuming more energy than necessary. With that, you have to take note of the proper temperature for the different parts of it, such as the freezer and the fresh food compartment. To check on their temperatures, just make use of a thermometer for it.

82. See that your refrigerator's door seals are tight.

You have to make sure that your refrigerator seals are really tight, so that cool air would not escape through it. One of the best things to do for it is to close the door while letting a paper hang half inside the unit and half outside of it. If you can pull the paper out easily, then it means that the door seals are not airtight.

83. Don't let your kids leave their TV sets turned on.

If your kids have their own rooms and each of the rooms has a TV set inside, then you should make sure that they are turned off when no one is using. With that, you should constantly remind your kids that they have to turn off the TV sets when they leave their rooms. Aside from that, make it a practice to constantly check on their rooms, just to make sure.

84. Make use of ceiling fans.

If you want to have a substitute for air conditioning, you can actually install ceiling fans inside your place for it. Ceiling fans can actually make a room more comfortable by providing sufficient movement of air. Because of that, you won't have to turn on your air conditioner a lot of times.

85. What to do with your guest room.

Having a guest room inside your house can increase your energy consumption, especially if you don't make adjustments when you don't need it. To conserve energy, see to it that the guest room's light is always turned off. Aside from that, keep in mind that attaining desirable temperature in it would not be needed as well.

86. Install tempered glass doors for your fireplace.

In having a fireplace, you want to make sure that heat loss would be reduced. This is because, if you can achieve that, it means that you would be able to reduce energy consumption of your heating equipment. With that, it is best to install tempered glass doors for your fireplace, so that heat loss is avoided, and efficiency is improved.

87. Small portable heaters.

Making use of portable heaters that are small in size is good if you use it for a shorter period of time. However, if you need heaters for the long haul, then it is best to make use of a baseboard heater for it. This type of heater is best for long term use as it consumes lesser amounts of energy.

88. It may be time to buy a new thermostat.

Always remember that as your thermostat becomes older, its efficiency would also reduce. With that, if you are using an old thermostat, then it may be time to replace it. By using a newer thermostat, you would be saving more energy, which would also help you save money in the long run.

89. Buying the right heating product.

When it comes to buying the right heating product, you should check its features first before making the final decision. Aside from checking on the features of the product, you should also check reviews about it. Reviews can be found online, which can help you decide when you are turned between two products.

90. Using your dishwasher.

It is always best to use your dishwasher with a full load. Doing it this way can help in conserving energy. Aside from that, keep in mind that using a dishwasher actually consumes lesser amounts of water as well as energy than washing the dishes manually. Thus, if you have enough dishes to wash, then use your dishwasher for it.

91. Use hot water only if it is really necessary.

One of the biggest factors that can make your electricity bills spike is the use of hot water. With that, it is best if you only use it if it is absolutely necessary. Thus, when it comes to washing your clothes, only use hot water when you are trying to clean extra dirty ones.

92. Adjust your washing machine properly.

If you are used to washing a lot of clothes with your washing machine, then you may always be setting it a higher levels. Keep in mind that doing this can make your equipment consume more energy. Aside from that, the said setting can also ruin delicate clothes. With that, it is always best to check on the equipment's settings, so that you won't be damaging your clothes and wasting energy.

93. Drying towels.

When you try to dry your towels with the use of your washing machine's dryer, then you would be using more energy. This is because, it would need to work harder, for towels and blankets can become heavy with water. To save energy, then it would be better to dry them on a separate load, or just simply air dry them.

94. Always read your oven's manufacturer's manual.

Keep in mind that electric ovens today have different operating features. In other words, there may be features that you can use in order to cook certain types of foods easier. With that, it is best to check on its manual, so that you can ensure that you won't be using more energy whenever you are cooking.

95. Try not to open the oven door.

Whenever you are cooking up something with your oven, you may be tempted to open its door from time to time. Keep in mind that whenever you open up the door, your oven's temperature could actually drop by 25 degree Fahrenheit. Thus, you would be making it work harder. Just check on the oven's timer instead or the clock for its progress.

96. Turn your drinking water heater off.

If you do not need to drink hot water, then you should turn off your drinking water heater. Keep in mind that heating the water constantly would consume a lot of energy. Even if you are not consuming hot water, as long as its turned on, it would still constantly consume energy.

97. Soak beans overnight.

If you are planning to cook beans in the morning for breakfast, then soak them overnight. This is because, soaking it would allow the beans to absorb water. Because of that, the time you would consume in cooking them would be reduced, as well as the energy consumed.

98. Boiling hot water.

When it comes to drinking coffee, you may want to boil water for it. With that, it is best to boil just enough hot water for your needs, so that you would conserve energy. Thus, if you only need a cup of coffee, then boil just a cup of water for it, so that you would save time and energy.

99. Turn off decorative indoor and outdoor lights.

If you have decorative indoor and outdoor lights, then you should turn them off, whenever you don't need them. Only use these types of lights when you are having a party or you are expecting guests. By doing that, you would not just save energy, but also prolong the lifespan of your lights.

100. Get the whole family involved.

It is best if you get your whole family involved in your quest to conserving energy. This is because, all of you would be consuming energy each day. Talk to them about the benefits of saving energy, as well as the consequences if you would not take action for it. Explain things properly, and make sure to make it clear to them on how you are going to go about it.

Conclusion

There seems to be more of an interest in the forms of renewable energy these days than ever before. People from all walks of life are seeing the many benefits it can offer. What should be an indicator that we need to continue moving forward is that many of the underdeveloped countries out there use more renewable energy than the rest of us.

While it is great that we have technology on our side, we need to always keep in mind what these products are doing to our environment. Global warming has always been a huge concern. The problem seems to continue to get worse though with the various types of pollution that are placed into the air.

This is believed to be a key factor in the strange things that go on around us with the weather. Desert areas have seen rain and snow in recent years. Areas that get substantial rainfall are now in a drought. Several natural disasters including hurricanes, floods, and tornadoes continue to destroy everything in their path.

As the governments out there as well as members of society continue to embrace what renewable energy has to offer it is likely that will start to come back into balance. There are plenty of major corporations out there leading the way as well. They want to set a very good example for others in the hopes that they will walk along the same path.

As the cost of fuel for our vehicles continues to increase everyone is worried about it. As a result it makes searching for an alternative form of renewable energy that can be used in place of it. Some of the vehicles out there known as hybrids do all this to help. They use solar power the majority of the time and only switch to gasoline as a backup until more solar energy can be collected.

By diversifying the various methods we use for renewable energy, we can balance out our desire to move forward as a society with the best of everything with our responsibility to protect the environment. There are many pros and cons to each of the renewable energy sources, but the benefits certainly outweigh the negativity.

There is no way around the fact that we can't replace the fossil fuel we consume. The more we do so the less that will be available for future generations. We can act on what we known and go with renewable energy or we can act selfish and continue to deplete what we have and leave the future generations to figure it out on their own.

To your success with a brand new green and clean world with renewable energy,

Albert Sicard

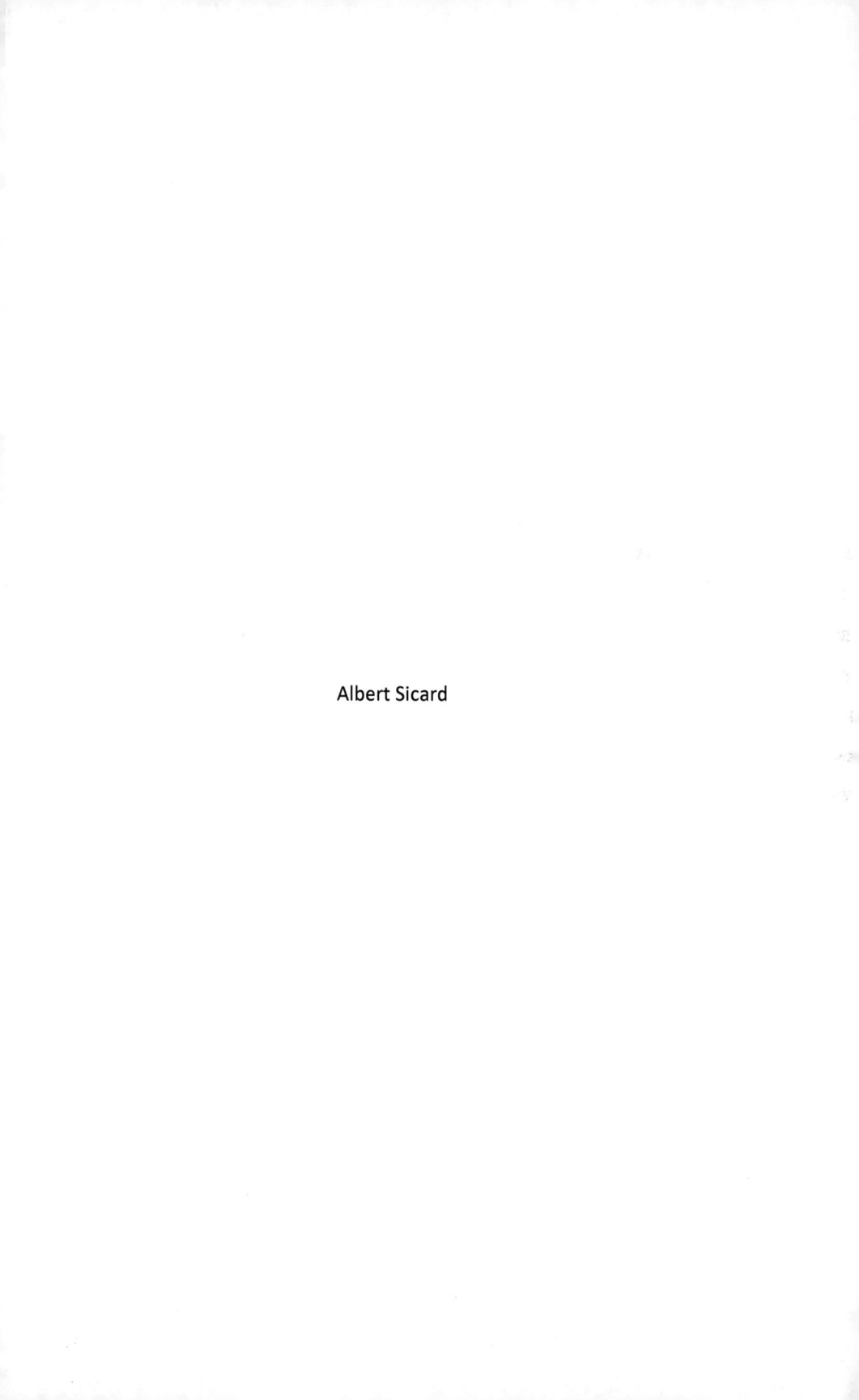

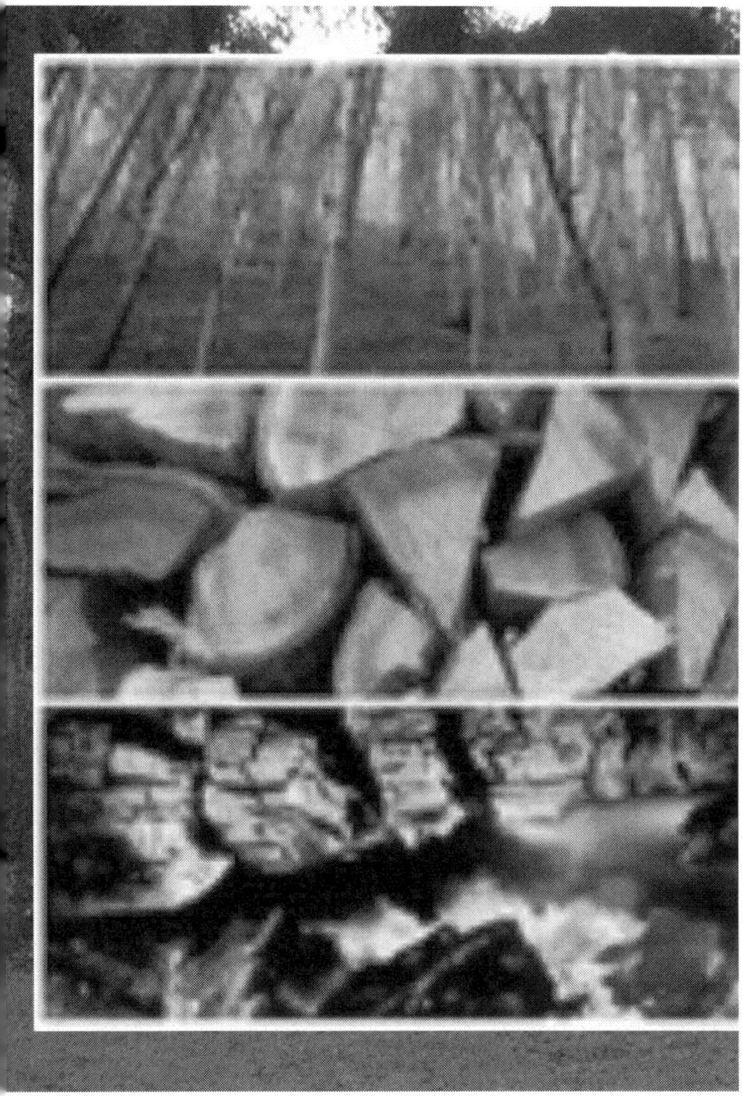

Convert Your Car To Burn Water In Addition To Gas And Save Up To 60%

Check it out here:

Resource 1

Run Your Car On Water, Make Biodiesel, & Save Money!

Check it out here:

Resource 2

Alternative Energy Resources

Check it out here:

Resource 3

Understanding and Installing Your Own Solar Electric System. Includes Worksheets,& Resources Keeping My Earth Green

You Can Learn How To Reduce Your Carbon Footprint And Help Reverse The Effects Of Global Warming

Check it out here:

Resource 4

Solar Power Design Manual

Teach Yourself All About Solar Power. Comprehensive Manual By Genuine Expert. Spreadsheet Included.

Check it out here:

Resource 5

Printed in Great Britain
by Amazon